MACHINE LEARNING UNLOCKED

Beginners' Guide to Learning and Understanding Machine Learning

Koso Brown

Contents

Introduction

Chatbots, predictive text, language translation apps, Netflix recommendations, and the layout of your social media feeds are all products of machine learning. It drives self-driving cars and devices that use photos to identify medical issues.

Machine learning is most likely being employed by businesses when they implement artificial intelligence systems these days; in fact, the terms are occasionally used ambiguously and interchangeably. A branch of artificial intelligence known as "machine learning" allows computers to learn without explicit programming.

Within the last five to ten years, machine learning has emerged as a crucial method—possibly the most significant—for completing the majority of AI tasks. That's why some people refer to machine learning and artificial intelligence practically interchangeably. The majority of AI advancements to date have included machine learning.

Everyone in business will probably come across machine learning because it is becoming more and more commonplace, so having a basic understanding of the subject is essential.

According to a 2020 Deloitte survey, 97% of businesses either currently utilize machine learning or plan to do so in the upcoming year. Of them, 67% employ it.

Even established businesses, including bakeries, banks, and retail stores, are utilizing machine learning to increase productivity and uncover new opportunities. Every industry is evolving or will change as a result of machine learning, and executives need to be aware of the fundamentals, the possibilities, and the constraints.

Although technical details are not necessary for everyone to grasp, people should be aware of what the technology is capable of and what it cannot.

This involves understanding the ethical, social, and societal ramifications of machine learning. It's critical to participate, get to know these technologies, and then consider how best to apply them.

Chapter 1: What is Machine Learning?

Artificial intelligence, roughly defined as a machine's ability to mimic intelligent human behavior, includes machine learning as a subfield. Complex tasks are carried out by artificial intelligence systems in a manner akin to human problem-solving.

AI aims to build computer models that display "intelligent behaviors" akin to those of humans, as stated by CSAIL's main research scientist and leader of the Info Lab Group Boris Katz. This includes devices that can identify an image, comprehend a document written in natural language, or carry out a task in the real world.

Using AI in machine learning is one method. AI pioneer Arthur Samuel defined it as "the field of study that gives computers the ability to learn without explicitly being programmed" in the 1950s.

According to Mickey Shulman, head of machine learning at Kenshoo, which specializes in artificial intelligence for the financial and U.S. intelligence communities, and lecturer at MIT Sloan, the term is accurate. He likened baking, in which a recipe specifies the precise quantity of materials to be used and the precise length of time to be mixed, to the conventional method of programming computers, or "software 1.0." In a similar vein, traditional programming calls for writing comprehensive instructions that the machine must obey.

However, there are situations (such as teaching a computer to distinguish between images of various persons) where developing a program for the machine to follow is impractical or takes a long time. It's easy for humans to perform this work, but teaching a computer how to do it is more challenging. Using an experience-based method, machine learning teaches computers how to program themselves.

Data—numbers, images, or text—is the foundation of machine learning. Examples of data include bank transactions, images of individuals or even specific bakery goods, repair records, time series data from sensors, and sales reports. To be utilized as training data, or the information the machine learning model will be trained on, the data is collected and prepared. The program is better the more data it has.

Programmers then select a machine learning model to utilize, provide the data, and allow the computer model to learn on its own to identify trends or anticipate future events. As time goes on, the model's parameters can be altered by a human programmer to assist it in providing increasingly accurate results.

Within machine learning, there are three subcategories:

1. **Supervised:** models for machine learning are fed labeled data sets during training, enabling the models to develop and become more precise over time. For instance, an algorithm may be taught with human-labeled images of dogs and other objects to teach the computer how to recognize dog images on its own. The most often utilized kind of machine learning nowadays is supervised learning.

2. **In unsupervised:** machine learning is a program that scans unlabeled data for patterns. Unsupervised machine learning is the ability to identify trends or patterns that humans aren't consciously searching for. An unsupervised machine learning software, for instance, may examine online sales data and recognize various customer types making purchases.

3. **Reinforcement:** By creating a reward structure, machine learning teaches machines to make the optimal decision through trial and error. By letting the machine know when it made the proper choices,

reinforcement learning can be used to teach models to play games or teach autonomous cars to drive. This allows the machine to gradually understand what actions are appropriate.

Additionally, machine learning is linked to the following additional subfields of artificial intelligence:

❖ Deep learning

Neural networks with multiple layers are called deep learning networks. In an image recognition system, for instance, certain layers of the neural network might detect individual features of a face, like eyes, noses, or mouths, while another layer would be able to determine whether those features appear in a way that indicates a face. This type of layered network is capable of processing large amounts of data and determining the "weight" of each link in the network.

Similar to neural networks, deep learning is based on the functioning of the human brain and is the basis for many applications of machine learning, including chatbots, driverless cars, and medical diagnostics.

Because deep learning demands large amounts of processing power, questions are raised regarding its viability from an economic and environmental standpoint.

❖ Natural language interpretation

In the discipline of machine learning known as natural language processing, computers are trained to comprehend human spoken and written language rather than the statistics and figures that are typically used to teach computers. This enables machines to produce new text, translate between languages, and recognize, comprehend, and respond to language. Familiar technology such as chatbots and digital assistants like Alexa or Siri are made possible by natural language processing.

❖ Neural networks

One popular, specialized type of machine learning technique is neural networks. The structure and interconnectivity of dozens or millions of processing nodes arranged into layers characterize the human brain, which is the model for artificial neural networks.

Each connected cell, or node, in an artificial neural network processes inputs and generates an output that is relayed to other neurons. Labeled data flows between the nodes, or cells, with a distinct function carried out by each cell. The many nodes in a neural network trained to determine whether or not an image features a cat would evaluate the data and produce an output that indicates whether or not a picture features a cat.

Applications of machine learning in business

Some businesses, like Netflix with its recommendation system and Google with its search engine, rely heavily on machine learning. Even though machine learning is not their primary business offering, other companies are actively involved in the field.

Others are still attempting to ascertain the most advantageous application of machine learning.

Researchers from the MIT Initiative on the Digital Economy published a study in 2018 that included a 21-question assessment for determining if a task is appropriate for machine learning. The researchers discovered that while no profession would be immune to machine learning, none is likely to be entirely replaced by it either. The researchers discovered that breaking down jobs into separate tasks—some of which can be completed by machines and others that need human intervention—was the key to unlocking the potential of machine learning.

Businesses currently employ machine learning in several ways, such as:

1. **Diagnostics and imaging in medicine.** Like a tool that can predict cancer risk based on a mammogram, machine learning programs can be trained to evaluate medical images or other data and look for specific indications of illness.

2. **Chatbots and automated hotlines.** A lot of businesses are using online chatbots, where clients or consumers converse with a machine rather than a human. These algorithms combine natural language processing and machine learning, and the bots learn from recorded conversations to provide relevant answers.

3. **Vehicles that drive themselves.** Machine learning, especially deep learning, is the foundation of a large portion of the technology used in self-driving automobiles.

4. **Fraud detection.** Machines can recognize possibly fraudulent credit card transactions, log-in attempts, or spam emails by analyzing trends, such as how an individual often spends their money or where they typically shop.

5. **Recommendation algorithms.** Machine learning powers the recommendation engines that power YouTube and Netflix suggestions, Facebook news feed content, and product recommendations. "They want to know, for example, what posts or liked content to share with us on Facebook, what ads to display, and what tweets we want them to show us on Twitter."

6. **Object detection and image analysis.** While face recognition techniques are debatable, machine learning is capable of analyzing photographs for a variety of information, including the ability to recognize and distinguish between people. This has a variety of business uses. Shulman pointed out that hedge funds are well-known for using machine learning to examine parking lot car counts to gain insight into businesses' performance and place-wise wagers.

Chapter 2: Machine Learning Engineers

Programmers with advanced technical skills, and machine learning engineers investigate, create, and construct self-executing software to automate prediction models. A machine learning (ML) engineer creates artificial intelligence (AI) systems that use massive data sets to create and develop learning algorithms that can eventually make predictions.

The software "learns" from the outcomes each time it runs, improving the accuracy of subsequent operations.

To help create high-performance machine learning models, machine learning engineers must evaluate, analyze, and organize data, run tests, and improve the learning process when designing machine learning systems.

What Is the Work of a Machine Learning Engineer?

Highly trained programmers known as machine learning engineers create artificial intelligence (AI) systems by researching, developing, and producing algorithms that can learn and make predictions using vast amounts of data.

In general, this position is in charge of creating machine learning systems, which entails gathering and analyzing data, carrying out experiments and testing, and generally keeping an eye on and improving machine learning procedures to create robust machine learning systems.

Knowledge of programming languages such as Python, Java, and C/C++ is required by many job requirements.

Job Description for a Machine Learning Engineer

All or most of the following functions are typically included in a job description for a machine learning engineer, however, specific duties will vary based on the size of an organization and the entire data science team:

1. Evaluating and prioritizing ML algorithm application cases based on the likelihood of success
2. Data visualization to gain a deeper understanding
3. Enhancing already-existing ML libraries and frameworks
4. Knowing when to use the results of your research to inform business decisions

5. Confirming data quality or making sure of it through data cleaning

6. Finding variations in the distribution of data that might have an impact on the performance of the model in practical settings

7. ML models and systems should be trained and retrained as necessary

8. Before beginning data collection and data modeling, look for and choose relevant data sets

9. Carrying out statistical analysis and utilizing the findings to enhance models

10. Studying, altering, and converting prototypes for data science

11. Creating, experimenting with, and studying machine learning models, techniques, and systems

Where Do Engineers Specialized in Machine Learning Originate?

While a machine learning engineer might begin their career in any number of fields, the majority of ML engineers have a background in data science, computer science, engineering, or mathematics.

The backgrounds of machine learning engineers and those in similar jobs, such as data scientists, software engineers, data analysts, and data engineers, differ, according to a study published in Indeed.

According to the survey, "Software Engineer" is the most likely previous job title for a Machine Learning Engineer, in terms of their professional history. Before pursuing a career in machine learning, a large number of other ML engineers had academic positions.

However, it's crucial to keep in mind that machine learning and data science are still relatively young disciplines of study. Additionally, as more tech companies and other businesses strive to expand their data science teams, more opportunities to become machine learning engineers are opening up.

While a strong foundation in math and computer science is still required, many people are gaining the additional knowledge and abilities needed to become Machine Learning Engineers by enrolling in certification courses, many of which can be completed online. These courses cover topics like supervised and unsupervised learning, deep learning, regression, classification, clustering techniques, and neural networks.

Chapter 3: Qualities of an Effective Machine Learning Engineer

1. They Have a Strong Intuition About Data

Without data analysis, machine learning cannot exist. A competent machine learning engineer or data scientist must be able to swiftly sort through huge data sets, spot trends, and understand how to use the information to draw conclusions that are useful and applicable.

They seem to have a sixth sense when it comes to data. Proficiency in data handling is essential.

They must be skilled in creating large data pipelines as well. It's also important to comprehend the power of vision. You need to be proficient with data visualization tools like Excel, Tableau, Power BI, Plotly, and Dash to guarantee that the insights you've uncovered are appropriately comprehended and valued by others.

2. They're Good Programmers of Computers

Learning to program is a prerequisite for pursuing a career in artificial intelligence or machine learning. Not only should a programmer be familiar with commonly used languages like Python, Java, and C++, but also other languages. Prolog, Lisp, and R are a few more languages that are becoming crucial for machine learning. All machine learning engineers, however, do not have to be proficiencies in HTML and JavaScript.

3. They Love the Iterative Process

Machine learning is an iterative process by definition. One needs to enjoy such a manner of development to be effective in this role. When creating a machine learning system, one must first quickly create a very basic model and then iterate through several stages to improve it.

Once more, though, an effective machine learning engineer must not be overly obstinate. It is important that you learn to recognize when to give up. Any machine learning system can always be made more accurate by iteratively improving upon it; the trick is to recognize when it's not worth the time and effort to continue.

4. They Have a Firm Basis in Statistics and Math

Without at least a basic understanding of math, machine learning is impossible to grasp. It won't matter if you have a formal foundation in math or statistics; to stay up, you must be proficient in the subject at least through high school. The foundation of numerous machine learning algorithms lies in a formal description of probability and methods that stem from it. The study of statistics, which offers a variety of measures, distributions, and analysis techniques required for creating and confirming models using observable data, is closely related to this. A lot of machine learning methods are essentially just statistical modeling techniques expanded upon.

5. Experts in machine learning are Innovative Problem Solvers

The most talented ML engineers are naturally curious. When a model or experiment fails, they are curious to know why rather than reacting with frustration.

However, they also effectively resolve issues. Because addressing specific faults will take time and make your models more complex and difficult to work with, the top machine learning experts create broad ways to fix bugs and misclassifications in their models.

It's also critical to strike a balance between your need to solve issues and your realistic awareness that many of your models and experiments will fall short. The most proficient Machine Learning Engineers acquire the ability to discern when to give up.

6. Skills for Machine Learning

You need to combine the knowledge and abilities of a data scientist and a software engineer to be successful as a machine learning engineer. This entails having a thorough understanding of all the core ideas in data analysis and computer science in addition to having certain soft skills related to both fields.

7. Skills in software engineering

Understanding data structures, being familiar with computer architecture, and writing algorithms—including those that can sort, optimize, and search—are some of the fundamental computer science ideas that ML engineers should be aware of. Because software is an ML engineer's usual product, they should also know how to adhere to best practices for software engineering, particularly in the areas of system architecture, version control, testing, and requirements analysis.

8. Data skills

Many of the skills that a data scientist and a machine learning engineer are required to possess are similar, such as data modeling, technical competency with programming languages like Python and Java, and the ability to assess prediction models and algorithms. Also, having a solid grasp of statistics and probability would be highly beneficial.

Is Coding Necessary for Machine Learning?

Yes, some coding experience is required if you want to work in the fields of artificial intelligence and machine learning. Because machine learning is implemented through code, programmers who are proficient in that language will have a solid understanding of how algorithms operate, making them better able to monitor and optimize those algorithms.

The three most common programming languages are Python, Java, and C++, while there are many more variations. When focusing on machine learning, it becomes crucial to understand languages like R, Lisp, and Prolog.

Ironically, one of the most fascinating aspects of machine learning is that its primary objective is to teach a computer to learn without the need for human assistance. To grasp the foundations of machine learning and to work with data matrices and vector operations with ease, you'll need to have a solid understanding of probability and statistics, complex linear algebra, and differential and integral calculus. Neural networks, supervised learning, unsupervised learning, reinforcement learning, and other fundamental machine learning paradigms are also crucial to understand.

Rather than beginning with coding and learning to write code, several Machine Learning Engineers advise anyone interested in pursuing the career to start with these fundamental ideas. It's imperative that you comprehend the fundamental ideas behind artificial intelligence.

Languages Used in Programming for Machine Learning

While there is some disagreement regarding the precise significance of coding skills for a machine learning engineer, it is generally accepted that to utilize, develop, and apply machine learning models and algorithms as efficiently as possible, one must acquire at least rudimentary programming skills.

These are the top programming languages for machine learning, whether you're seeking to pursue a career in the subject or want to increase your knowledge of it to work in data science or analytics:

1. Julia

Julia is a more niche and recent programming language that is particularly well-liked for high-performance numerical analysis, machine learning model analytics, and creating machine learning applications. Julia is a fast-programming language with universally executable code that can be compiled in Julia from Python or R via a wrapper, and it supports a wide range of hardware.

Julia's ecosystem provides some helpful machine-learning packages. Popular machine learning techniques such as clustering, decision trees, and generalized linear models have a single interface provided by the MLJ.jl package; Turing. jl is a potent package for probabilistic programming, and Flux. jl is helpful for deep learning.

2. Java and JavaScript

Python, JavaScript, and Java are multifunctional programming languages with a seemingly limitless array of potential uses, including machine learning. Regression techniques, data processing, and the development of machine learning algorithms may be accomplished quickly and effectively with Java and JavaScript.

Data pretreatment, preparation, clustering, and classification are just a few of the machine learning and data mining operations that are supported by Java frameworks like Weka, Rapid Miner, and JavaML. In the meantime, one can construct and train an ML model or incorporate neural networks using JavaScript tools.

3. R Programming Language

R is a well-liked open-source programming language that is frequently used in statistical computing. Data scientists who work with huge amounts of statistical data frequently choose to use R. Many machine learning applications, such as data sampling, data visualization, supervised or unsupervised learning, and training machine learning models, can be accomplished with R.

R, which also has an extensive library of helpful packages, is frequently used to apply machine learning techniques including regression, decision tree construction, and classification.

4. C/C++

Machine learning has shown these programming languages to be effective and valuable due to their versatility, power, and speed. You can adjust a wide range of performance factors or work with machine learning methods using C++. Strong open-source libraries for C++, such as mlpack, Torch, and TensorFlow, can execute numerical and scientific operations, increase productivity, and provide a wide range of well-liked machine learning methods.

5. Python Programming Language

One of the world's most popular programming languages with near-endless applications in data science, data analysis, artificial intelligence, web development, and software engineering, Python has also become one of the most important languages for machine learning because of its readable code, flexibility, and a vast collection of libraries and packages.

For many intricate machine learning applications and frameworks, the use of Python libraries and packages might be essential in terms of time and effort savings. For example, sci-kit-learn is great for building ML algorithms, NumPy can be the finest tool for working with textual data, and TensorFlow or Keras is good for deep learning.

6. Career Path

Certainly, a career in machine learning is rewarding. In terms of pay, the number of job listings growing, and overall demand, machine learning engineering is one of the most sought-after careers in the US, according to a recent Indeed research.

The basic compensation for Machine Learning Engineers has risen to almost $140,000, while the number of job listings has surged by approximately 350 percent since 2015.

If you have a strong interest in data science, automation, and algorithms, a career in machine learning could be ideal for you. Moving massive volumes of raw data, applying machine learning algorithms to process that data, and then automating the process for optimization will occupy your days.

An additional factor that makes a profession in machine learning so alluring? Professionals in the machine learning field have an abundance of employment options to select from. You can find well-paying positions as a Machine Learning Engineer, Data Scientist, NLP Scientist, Business Intelligence Developer, or Human-Centered Machine Learning Designer if you have experience with machine learning.

The high demand and scarcity of individuals possessing machine learning skills contribute to the high salaries of these occupations. Even tales of bidding wars for artificial intelligence (AI) talent exist, as big firms scramble to sign the brightest minds in the field.

Career Paths for Machine Learning Engineers

Not many IT workers start as machine learning engineers. The majority of people who eventually go into machine learning careers do so from positions such as data scientist, data engineer, software engineer, software programmer, or developer.

The two most popular career pathways for becoming a machine learning engineer are usually in data or software development, albeit the latter does call for a good deal of coding language skills. To hone their coding skills, a machine learning expert with development experience could pursue computer science coursework or enroll in a coding boot camp.

After working in the field for a few years, a person may decide to specialize in deep learning, automation, cloud computing, data, or other specialty roles, or they may begin applying for senior-level machine learning positions.

For example, there is currently a great need for tech workers with experience in natural language processing; these workers should search for positions such as NLP Scientist or NLP Engineer. Another position that may be of interest to someone with advanced machine learning expertise is Human-Centered Machine Learning Designer.

Chapter 4: Benefits of Machine Learning

- **Cut unplanned timeline**
- **Detect Fraud**
- **Increase Revenue**
- **Boost efficiency**
- **Retain customer**
- **Improve planning and forecasting**
- **Addresses industry needs**

Types of machine learning algorithms

1. **AdaBoost.** This supervised learning method, also known as adaptive boosting, strengthens a poor ML classification or regression algorithm by mixing it with stronger ones to create a stronger algorithm that generates fewer errors.

2. **Gradient boosting.** By using this optimization technique, the cost function of a neural network—a measurement of the amount of error the network generates when its actual output differs from its anticipated output—is lowered.

3. **Dimensionality reduction.** A data set is said to have high dimensionality if it contains a large number of features. Reducing the number of characteristics to retain only the most significant insights or data is referred to as dimensionality reduction. Principal component analysis is one instance of this technique.

4. **Networks of artificial neurons.** ANNs, also known as neural networks, are collections of algorithms that use neurons as building blocks to identify patterns in incoming data. These neurons are somewhat similar to brain neurons in humans. Through supervised training techniques, they are gradually trained and changed.

5. **K-nearest neighbors.** KNNs use similarity or proximity to classify data items. The new data piece will be grouped with the existing data group that most closely resembles it.

6. **K-means.** Within unlabeled data sets, our unsupervised learning technique finds data categories. As one of the most often used clustering algorithms, it divides the unlabeled data into several clusters.

7. **Random forest.** Regression and classification techniques are used by these algorithms to organize and

categorize data by combining several unconnected decision trees.

8. **Naïve Bayes.** Predictions and classifications are carried out by this algorithm. Being one of the most basic supervised learning algorithms, it assumes that every characteristic in the input data is independent of every other data point, meaning that when it comes to producing predictions, one won't have an impact on another.

9. **Support vector machine.** Regression analysis, anomaly detection, and categorization of data are all done with SVMs. When items from a data set are divided into two separate categories, known as binary classifications, an SVM performs best.

10. **Decision tree.** This approach for supervised learning is applied to problems related to regression as well as classification. Decision trees use a set of questions or conditions to identify which subset each data element belongs in, thereby dividing data sets into multiple subsets. The term "tree" refers to the way that data seems to be divided into branches when it is mapped out.

11. **Linear regression.** A supervised technique called a linear regression algorithm is used to forecast continuous numerical values that vary or fluctuate over time. Over time, it can learn to predict variables with more accuracy, such as age or sales figures.

12. **Logistic regression.** A machine learning algorithm is usually a component of predictive modeling in predictive analytics, which employs historical data and observations to forecast the likelihood of future occurrences. Similar to supervised algorithms, logistic regressions concentrate on binary classifications as results, such as "yes" or "no."

Why Should We Learn Machine Learning?

A valuable technology that may be applied to many different types of challenges is machine learning. It enables computers to learn without explicit programming by using data. This enables the development of systems that, by learning from their experiences, can automatically perform better over time.

There are many reasons why learning machine learning is important:

1. The discipline of machine learning is expanding quickly and offers a wealth of interesting research and

development opportunities. Gaining knowledge in machine learning will enable you to remain current with the most recent findings and advancements in the field.

2. An essential tool for data analysis and visualization is machine learning. It enables you to draw conclusions and trends from huge datasets, which are useful for comprehending intricate systems and formulating wise choices.

3. Numerous sectors, including e-commerce, finance, and healthcare, heavily rely on machine learning. You can pursue a variety of job options in these domains by learning machine learning.

4. Intelligent systems that can make decisions and predictions based on data can be created via machine learning. This can assist businesses in improving their operations, coming up with new goods and services, and making smarter judgments.

Chapter 5: How to get started with Machine Learning

Let's start by reviewing some of the key terminology.

Terminology:

- **Underfitting:** It is the situation in which the input data's underlying trend cannot be discerned by the model. It completely undermines the machine learning model's accuracy. To put it simply, not enough of the model or algorithm fits the data.

- **Overfitting**: A machine learning model that has been trained on vast amounts of data often learns from noise and incorrect data entry. In this case, the model does not accurately characterize the data.

- **Prediction:** When the machine learning model is prepared, input data may be fed into it to produce an output prediction.

- **Model:** A machine learning model, also called a "hypothesis," is a mathematical depiction of a real-world procedure. A machine learning model is created by combining a machine learning algorithm with training data.

- **Feature Vector**: It is a collection of several numerical attributes. For training and prediction, we feed it into the machine learning model.

- **Feature:** A measurable attribute or parameter of the data set is called a feature.

- **Target (Label):** The target or label refers to the value that the machine learning model needs to forecast.

- **Training:** An algorithm's input is a collection of data referred to as "training data." By identifying patterns in the input data, the learning algorithm trains the model to produce the desired outcomes. The machine learning model is the result of the training process.

Steps to Machine Learning

There Are Seven Machine Learning Steps.

1. Gathering Data
2. Preparing that data
3. Choosing a model
4. Training
5. Evaluation
6. Hyperparameter Tuning
7. Prediction

Learning a programming language—Python is preferred—as well as the necessary analytical and quantitative skills are prerequisites. Before tackling Machine Learning problems, you should review the following five mathematical topics:

- ❖ Complex Optimizations and Algorithms
- ❖ Multivariate Calculus
- ❖ Mathematical Analysis: Gradients and Derivatives
- ❖ Scalars, Vectors, Matrices, and Tensors in linear algebra for data analysis
- ❖ Machine learning using statistics and probability theory

Distinctive Features of Artificial Intelligence, Deep Learning, and Machine Learning

Concept	Definition
Artificial intelligence	**The goal of computer science is to build intelligent machines with human-like thought and behavior.**
Machine learning	**A branch of artificial intelligence that specializes**

	in creating models and algorithms that may be deliberately taught to learn from data.
Deep learning	A branch of machine learning that recognizes complicated patterns in data by training multi-layered artificial neural networks.

The key distinctions between these ideas are briefly summarized as follows:

- ❖ The creation of intelligent systems is the focus of the large field of artificial intelligence, which includes many different methods and strategies.
- ❖ Machine learning, a branch of artificial intelligence, is the process of teaching algorithms to learn from data instead of directly programming them.
- ❖ A subfield of machine learning called "deep learning" uses several layers of artificial neural networks to find complex patterns in data.

Benefits and Drawbacks of Machine Learning

There are benefits and drawbacks to everything. In this section, we will discuss some of the fundamental benefits and drawbacks of machine learning.

Benefits:

1. It can be used to automatically find data outliers.
2. It can be applied to automatically cluster data.
3. It is useful for automatically creating new features from data.
4. Forecasts on upcoming data can be made using it.
5. Finding patterns with it is possible.

Drawbacks:

Overfitting, explainability issues, and the possibility of skewed data are among the drawbacks.

Tools for Machine Learning

Machine learning engineers should be proficient in Python, Java, and C++ coding and development, but many additionally find it beneficial to become knowledgeable about the following machine learning tools and resources:

Amazon Machine Learning

- ❖ TensorFlow
- ❖ Spark and Hadoop
- ❖ Google Cloud ML Engine
- ❖ Apache Kafka
- ❖ R Programming
- ❖ MATLAB

Conclusion

It is difficult to predict machine learning's exact future because it is a field that is always evolving and influenced by a wide range of factors. However, it seems likely that machine learning will remain a significant force in many areas of science, technology, and culture, as well as a major driver of technological innovation. Machine learning may be used in the future to create self-driving cars, intelligent assistants, and individualized healthcare. Machine learning has the potential to address significant global concerns such as poverty and climate change.